Theater Air Base Resiliency Optimization Model (TAB-ROM)

Analyst Guide for Version 2.0

RACHEL COSTELLO, ELI M. KOHLENBERG, BRENT THOMAS,
BRYCE DOWNING, JACOB L. HEIM, SASCHA ISHIKAWA

Prepared for the Department of the Air Force
Approved for public release; distribution is unlimited.

RAND PROJECT AIR FORCE

For more information on this publication, visit **www.rand.org/t/TLA3060-1**.

Published by the RAND Corporation, Santa Monica, Calif.

Library of Congress Cataloging-in-Publication Data is available for this publication.

ISBN: 978-1-9774-1524-0

Cover: Yasuo Osakabe/U.S. Air Force.

About This Tool

RAND researchers, working on RAND's Combat Operations in Denied Environments (CODE) series of projects, have developed a suite of integrated models that assess the vulnerability of land-based aircraft and air bases to missile attacks. Through CODE projects, RAND researchers identify cost-effective mixes of investments in air base resilience, gauge the effects of prepositioning strategies on force arrival times and sustainment costs, and suggest how to improve an airfield's infrastructure and force protection.

This analyst guide details the inputs, algorithms, and outputs of the Theater Air Base Resiliency Optimization Model (TAB-ROM). The guide reflects the state of TAB-ROM 2.0 as of September 2023 and is intended as a companion to in-person training of analysts in the U.S. Air Force (USAF), the Office of the Secretary of Defense (OSD), or research communities who use the Theater Air Base Vulnerability Assessment Model (TAB-VAM) and TAB-ROM to evaluate specific operational plans and scenarios. TAB-ROM is a product of research sponsored by the USAF and OSD over many years.

RAND Project AIR FORCE

RAND Project AIR FORCE (PAF), a division of RAND, is the Department of the Air Force's (DAF's) federally funded research and development center for studies and analyses, supporting both USAF and the United States Space Force. PAF provides the DAF with independent analyses of policy alternatives affecting the development, employment, combat readiness, and support of current and future air, space, and cyber forces. Research is conducted in four programs: Strategy and Doctrine; Force Modernization and Employment; Resource Management; and Workforce, Development, and Health. The research reported here was prepared under contract FA7014-22-D-0001.

Additional information about PAF is available on our website:
www.rand.org/paf/

The information in this tool does not reflect the official policy or position of the Department of Defense or the U.S. Government.

Funding

Funding for this research was made possible through the concept formulation provision of the DAF–RAND Sponsoring Agreement. PAF uses concept formulation funding to support a variety of activities, including research plan development; direct assistance on short-term, decision-focused DAF requests; exploratory research; outreach and communications initiatives; and other efforts undertaken in support of the development, execution, management, and reporting of PAF's approved research agenda.

Acknowledgments

This work would not have been possible without the support from many people from a wide variety of communities within the USAF and OSD who contributed to the development of the CODE analytic framework, of which TAB-ROM is a core component.

At RAND, we first thank Robert Tripp. His vision and leadership brought about the CODE analytic framework, which helps the USAF and OSD understand the extent of threats to U.S. and allied air operations in specific scenarios and to evaluate the cost-effectiveness of a wide variety of mitigation actions. His efforts were instrumental in bringing about the CODE analytic framework, without which TAB-ROM would not exist.

We also thank Andrew Karode. As the lead software engineer for the first decade of the CODE series of projects, his contributions underpin all CODE analysis. He led the effort to rebuild the original version of TAB-ROM into the form described in this guide, and he led much of the analysis performed with it in recent years. We would also like to thank Ricardo Sanchez and Trung Tran for their work in developing TAB-ROM.

We are grateful to Christopher Lynch, Daniel Romano, and Katherine Pfrommer for their leadership and analysis that led to leveraging and improving TAB-ROM. Their contributions have had a significant effect on the success of the CODE approach. We would also like to thank RAND's research management team, past and present, for their guidance and support, particularly Stephanie Young, Christopher Mouton, Obaid Younossi, Joel Predd, Cynthia Cook, and Laura Baldwin. We thank Kate Giglio for her help writing this guide, which improved its clarity and readability.

Finally, we would like to offer special thanks to RAND alumnus Harry Markowitz. His practical optimization approaches to model and manage portfolio risk were foundational in the approaches we present in this document.

Although many people contributed to the work described here, sole responsibility for its content belongs to the authors.

Contents

Figures and Tables

Figures

Tables

Chapter 1

Introduction

RAND researchers, working on RAND's Combat Operations in Denied Environments (CODE) series of projects, have developed a suite of integrated models that assess the vulnerability of land-based aircraft and air bases to missile attacks.[1] CODE projects identified cost-effective mixes of investments in air base resilience, gauged the effects of prepositioning strategies on force arrival times and sustainment costs, and suggested how to improve an airfield's infrastructure and force protection.

At the heart of this analytic framework is the Theater Air Base Vulnerability Assessment Model (TAB-VAM), which assesses the effect of missile attacks on air bases. TAB-VAM models adversary missile attacks and computes the resulting damage to critical assets including runways, aircraft parked in the open or in shelters, fuel infrastructure, munitions storage, aircraft spares and maintenance support equipment, barracks, and ground-based active missile defenses.[2] The tool takes into account protection measures, resupply rates, and runway recovery times and predicts the effect of attacks on aircraft sortie generation.[3] TAB-VAM accounts for uncertainty in an adversary's attack preferences by simulating a wide variety of attack vectors. This approach is crucial because in a multimove competition, the shifting incentives for different attack vectors can necessitate different mitigations as well.

The Theater Air Base Resiliency Optimization Model (TAB-ROM), a companion model to TAB-VAM, seeks to identify the most cost-effective mitigation options. TAB-ROM employs a genetic algorithm (GA), or computer algorithm based on evolution, to identify the most promising investments. It runs TAB-VAM tens of thousands of times, exploring the universe of possibilities for investments that show promise against dozens of plausible attack vectors, then narrows down the investments to those that appear to be the most cost-effective (e.g., solutions that yield the highest sortie rates at the lowest costs). TAB-ROM calculates the maximum number of sorties that can be generated at various spending levels, starting with no additional investments in air base resilience. The

---

[1] For a brief overview of the major findings from RAND's CODE work, see Christopher Lynch, Rachel Costello, Jacob L. Heim, Andrew Karode, Patrick Mills, Robert S. Tripp, and Alan J. Vick, *Operational Imperative: Investing Wisely to Bolster U.S. Air Bases Against Chinese and Russian Attacks*, RAND Corporation, PE-A1996-1, January 2023. For a description of the CODE methodology, see Brent Thomas, Mahyar A. Amouzegar, Rachel Costello, Robert A. Guffey, Andrew Karode, Christopher Lynch, Kristin F. Lynch, Ken Munson, Chad J. R. Ohlandt, Daniel M. Romano, Ricardo Sanchez, Robert S. Tripp, and Joseph Vesely, *Project AIR FORCE Modeling Capabilities for Support of Combat Operations in Denied Environments*, RAND Corporation, RR-427-AF, 2015.

[2] A TAB-VAM user does not need to include all of these assets in their modeling. Any combination of these assets can be included in the sortie generation calculation or excluded depending on the needs of the research questions. Excluding assets decreases the input burden on analysts and reduces run times.

[3] For example, TAB-VAM has been used to model the benefits of various forms of runway repair, aircraft shelters, redundant fuel storage, dispersed munitions storage, dispersed maintenance equipment and spares storage, personnel shelters, additional medical capabilities, and various ground-based active defenses.

results can be used to identify the "knee in the curve"—that is, an investment level at which point any higher investments would yield diminishing returns.

This analyst guide details the inputs, algorithms, and outputs of TAB-ROM. It reflects the state of TAB-ROM 2.0 as of September 2023. It is intended as a companion to in-person training of analysts in the U.S. Air Force (USAF), the Office of the Secretary of Defense (OSD), or research communities who use TAB-VAM and TAB-ROM to evaluate specific operational plans and scenarios.

This guide assumes familiarity with TAB-VAM, a Monte Carlo simulation model that analyzes complex trade-offs between basing and threat-mitigation strategies. TAB-VAM will provide much of the input necessary to run TAB-ROM. Instructions for TAB-VAM are not provided in this guide.

The CODE Analytic Framework

RAND's CODE framework has been used in many air base resilience projects. CODE consists of a suite of integrated models that have been implemented to assist analysts in numerous ways that include the following: (1) assessing the vulnerability of land-based U.S. and allied aircraft to missile attacks, (2) identifying cost-effective mixes of investments in air base resilience, (3) gauging the effects of prepositioning strategies on force arrival times and sustainment costs, and (4) suggesting how to improve an airfield's infrastructure and force protection (for example, with security forces and reorganized base defenses). There are eight RAND-developed tools used in CODE:

- **Global Airfield Database (GAD):** A database of global airfields, particularly the geographical and physical attributes of airfields, used to assess vulnerability to kinetic attack. GAD draws on multiple data sources; rationalizes, reconciles, and documents data; and synthesizes the data in a single, web-based database application that can be queried and can output data files for use by analysts and by a variety of models.
- **Theater Air Base Vulnerability Assessment Model (TAB-VAM):** A tool to model the effects of missile attacks on operations at air bases and to evaluate potential options to minimize damage and casualties and improve sortie generation under attack.
- **Ground Forces Vulnerability Assessment Model (GF-VAM):** Originally developed from TAB-VAM, GF-VAM models kinetic attacks on supply chains and logistics infrastructure and the effects of those attacks on sustainment and combat power.
- **Aimpoint Optimization (APO):** A tool used to identify the most-efficient Red aimpoints to maximize parking ramp coverage with a given number and mix of missiles. APO produces inputs for TAB-VAM. It can also be used independently to assess the overall vulnerability of air bases to parking attacks or to generate lookup tables for rapid wargame adjudication.
- **Theater Air Base Resiliency Optimization Model (TAB-ROM):** An optimization model that produces the Pareto frontier of cost-effective mitigation portfolios to improve the resiliency of air bases to missile attack.[4] TAB-ROM runs a TAB-VAM scenario tens of thousands of times, changing the active defense, passive defense, and infrastructure

[4] The concept of a *Pareto frontier* is discussed in more detail in Chapter 3.

investments with each run. The object is to maximize sortie generation in the face of adversary attacks (represented by a range of attack vectors) while minimizing cost.

- **Asset Placement Tool (APT):** A tool to add infrastructure and other assets to an existing airfield shapefile. The tool determines the number of aircraft or assets that can fit on a base subject to user-defined restrictions. It can produce multiple laydowns on multiple locations to allow analysts to explore possibilities (or to produce inputs for other tools, such as TAB-VAM and TAB-ROM).
- **Lean Strategic Tool for the Analysis of Required Transportation (Lean-START):** A tool to estimate the required personnel and equipment to support a USAF operating location, defined by such user inputs as type and number of aircraft, operational duration, available and desired levels of capability, and host nation support.
- **Prepositioning Requirements Optimization (PRePO):** An optimization model designed to evaluate given storage postures or identify optimal storage postures (with respect to such metrics as minimum cost, closure time, or transportation utilization) while meeting a set of specified constraints (e.g., storage and throughput capacity, loadout restrictions, transportation time requirements) for a given scenario.

Table 1.1 shows how the CODE models work together and the possible inputs and outputs for them.

Table 1.1. The CODE Analytic Framework

Inputs	Models	Outputs
Force postures • U.S. and allied bases and force beddowns • U.S. and allied concepts of operation (CONOPS) • GAD geospatial infrastructure for airfields • Adversary attack capabilities (missile ranges, warheads, and launch points) and attack vectors (missiles targeting parked aircraft, runways, fuel storage, munitions storage, etc.)	• TAB-VAM • GF-VAM	• U.S. and allied sortie generation under attack • Damaged aircraft (fighters, bombers, tankers, etc.) • Available runways • Destroyed and consumed fuel • Damaged missile defense systems • Personnel casualties (killed, wounded, evacuated, died of wounds, returned to duty)
Geospatial target data • Location geometry data for air bases, ports, army staging areas	• APO	• Adversary missile allocations
Geospatial constraints • A geospatial representation of existing assets and infrastructure at each location • Asset dimensions • Separation distances	• APT	• Location- and asset-specific constraints

Inputs	Models	Outputs
Resiliency options and costs • TAB-VAM scenario • TAB-VAM force posture • Active defense options • Passive defense options • Infrastructure options • Recovery capabilities • Location- and asset-specific constraints	• TAB-ROM	• Portfolio of active defenses, deployable passive defenses, and infrastructure options • Pareto frontier of optimal mitigation portfolios across the cost curve • Sortie generation under attack • Location-specific investments for each portfolio
Combat support requirements • Existing capabilities of operating locations • Aircraft and missions assigned to each location	• Lean-START	• Combat support requirements • Medical requirements
Logistics requirements and costs • Combat support, medical, and passive defense requirements • Transportation options • Storage options	• PRePO	• Prepositioning options

NOTE: In this context, the Pareto frontier is the set of mitigation portfolios that provides the most sortie generation for a given set of investment levels.

Although Table 1.1 shows a linear process across rows, many outputs from one model can be fed into another model as inputs in a continuous feedback loop. Furthermore, each model can be used alone or in combination with others in the framework. For some of the models, analysts may "turn off" certain features to simplify the analysis.[5]

This analytic framework has been a key part of CODE analyses that have informed USAF and OSD decisionmaking and operational planning since 2011. For example, as of 2023, OSD, the Joint Staff, and Headquarters Air Force have begun to invest in a passive defense portfolio to improve operational resiliency in the Pacific theater. As the scope of CODE analyses increases, RAND researchers continue to expand the analytic framework to include a broader range of force posture options, additional classes of assets that might come under attack, and an ever-widening range of materiel and nonmateriel mitigation options to test.

The Theater Air Base Resiliency Optimization Model

TAB-ROM is a GA written in Java that optimizes a mitigation portfolio of active and passive defenses. It can trade between active defenses (e.g., Patriot and Terminal High Altitude Area Defense missile systems), passive defenses (e.g., aircraft shelters, fuel bladders), recovery capabilities (e.g., runway repair), and air base infrastructure (e.g., additional runways, munitions bunkers) to build

[5] For example, TAB-VAM will always output sorties generated under attack; however, analysts could choose to ignore personnel casualties, which omits the need to include base populations and medical capabilities.

portfolios of mitigation options that offer the largest increases in theaterwide sortie generation in the face of an adversary's attack given the existing infrastructure and space constraints at each operating location. TAB-ROM uses a particular type of GA, the Nondominated Sorting Genetic Algorithm–II (NSGA-II),[6] to execute a multiobjective optimization of the mitigation portfolio. Currently, TAB-ROM has two objective functions: maximizing sortie generation while minimizing cost and accounting for practical constraints.

As inputs, TAB-ROM requires an entire TAB-VAM scenario. The scenario includes all inputs and the TAB-VAM executable file, a set of adversary attack vectors, the set of potential investment choices (which must be defined in TAB-VAM) along with the choices' costs, the maximum numbers of each investment choice allowed at each location, and, optionally, the specific location on the base that each investment choice resides (which APT can provide).

Given this list of inputs, TAB-ROM will test candidate mitigation portfolios by running each adversary attack vector against the TAB-VAM scenario to determine the effectiveness of the portfolio. Ultimately, this means that TAB-ROM might run TAB-VAM upward of a few hundred thousand times to generate the Pareto frontier of investment portfolios at different costs. Figure 1.1 shows the relationship between the portion of the CODE analytic framework required to run TAB-ROM.

Figure 1.1. The Portion of the CODE Analytic Framework Required to Run TAB-ROM

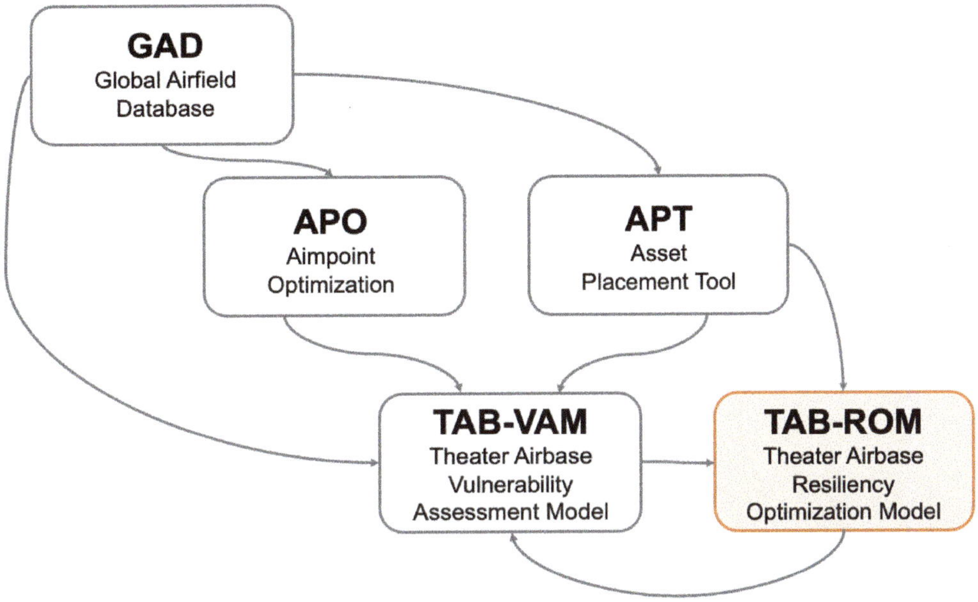

For each portfolio along the Pareto frontier, TAB-ROM provides what was purchased, the portfolio's total cost, and the resulting sortie generation. Analysts can then view those portfolios and determine which investments were driving the results, which investments might be interchangeable, and which investments do not enter the solution until very high price points. This allows the analyst to better understand the tradespace of potential resiliency investments.

---

[6] Kalyanmoy Deb, Amrit Pratap, Sameer Agarwal, and T. Meyarivan, "A Fast and Elitist Multiobjective Genetic Algorithm: NSGA-II," *IEEE Transactions on Evolutionary Computation*, Vol. 6, No. 2, April 2002.

Scope and Organization of This Analyst Guide

This guide presents an overview of TAB-ROM algorithms, inputs, and outputs. Chapter 2 describes the model's algorithms, including the objective functions, constraints, and the GA. Chapter 3 describes the inputs required to run TAB-ROM. Chapter 4 describes the output tables that TAB-ROM generates.

Important terms specific to CODE and its models, including TAB-ROM and TAB-VAM, are defined throughout the document. They are italicized in the text and briefly defined in text boxes. For analysts seeking more nuanced definitions, we recommend referencing the other CODE-related guides and studies referenced throughout the work.

This guide is supported by three appendices. Appendix A offers details about the input files described in Chapter 3 and defines all variables used by TAB-ROM. Appendix B describes each output table described in Chapter 4 in detail and defines the columns in each. Finally, Appendix C includes instructions for installing and running TAB-ROM.

Chapter 2

Algorithms

The CODE analytic framework was initially developed to answer one seemingly simple question: Given a menu of options to choose from, which ones should the U.S. Department of Defense invest in to improve the resiliency of air base operations to missile attacks? More specifically, the CODE framework was designed to address the types of investments that should be made at different locations to best improve the ability of U.S. land-based, fixed-wing aircraft to survive attacks on the ground and continue to fly missions in a contested environment.

The investment options that the Department of Defense can choose from vary in terms of the aspects of air base operations they protect (e.g., refueling, rearming, runways) and how they improve resiliency (e.g., redundancy, hardening, recovery, active defense). To make trade-offs between such disparate capabilities requires a common metric for comparison. The CODE framework uses the primary metric of *percent sortie generation*, which is the percentage of planned sorties achieved as defined in the box to the right.[7] (In this chapter, key terms are defined or given more information in callout boxes, such as the one to the right.) In our work, we typically count sorties from all operational aircraft in theater, spread across many operating locations, over a specified period. We developed TAB-VAM as a Monte Carlo simulation model to evaluate how well candidate investment options improve operational sortie generation during this period in the face of a variety of adversary attack strategies.

> **Percent sortie generation:** the number of aircraft that are able to launch after an attack divided by the number of planned launches before the attack, as specified by the user

TAB-ROM, when paired with TAB-VAM, is used to assess which combinations of investments are the most cost-effective (i.e., which investments provide maximum sortie generation at minimum cost) for a range of budget sizes. TAB-ROM's outputs are used to identify the set of investments, by location, for a given budget. Usually, the benefits gained from even relatively small investments compared with no investment are quite large, but the marginal gains in operational benefits decrease as the total investment amount increases. By identifying the investment levels that produce diminishing returns, TAB-ROM also provides insight into how much money is worth investing in the menu of capabilities under consideration within the context of a given scenario. TAB-ROM helps answer USAF decisionmakers' most typical questions: "What mix of investments will give me the best bang for my buck when it comes to air base defense?" and "What return on investment can I expect if I invest a given amount of money into air base defense?"

---

[7] TAB-VAM provides a fair bit of detail beyond sortie generation, including damage to aircraft on the ground, damage to infrastructure, aerial refueling demands, and many other metrics. For each metric, users can focus on specific aircraft types, such as tanker sortie generation or tankers damaged. Similarly, TAB-ROM can optimize other metrics calculated by TAB-VAM; however, no CODE project to date has used TAB-ROM in this manner.

In this chapter, we will describe the algorithms TAB-ROM uses to make these determinations. We begin with a brief reminder about TAB-VAM's capabilities.

The Theater Air Base Vulnerability Model

Because running TAB-ROM requires an entire TAB-VAM scenario, an analyst will need to understand TAB-VAM to understand TAB-ROM.[8] For those needing only review, the following are a few aspects of TAB-VAM to keep in mind:

- **Blue inputs.** There are three key U.S. (Blue) inputs to TAB-VAM, as follows:
  - detailed information on the air bases under consideration, including each air base's name, location (latitude and longitude coordinates), a geospatial representation of the air base's existing infrastructure (i.e., runways, aircraft parking areas, aircraft shelters, fuel storage, munitions storage, spares and maintenance equipment, billeting, and personnel shelters), assumed resupply rates (e.g., for fuel and munitions), medical capabilities, and base population
  - the aircraft laydown, along with the planned sortie rate, assigned mission, mission locations, and munitions loadout (if applicable) for each squadron
  - aircraft planning factors, including fuel capacity, fuel burn rates, and the length of runway needed for takeoff and landing.[9]
- **Red inputs.** The key adversary (Red) input to TAB-VAM include the set of missiles to be fired at air bases during the model run and the launch points, range, accuracy, and warheads available for each missile. The user must specify permissible missile-target combinations (i.e., Blue assets) and the corresponding degree of damage should the missile hit its target. TAB-VAM's targeting algorithm will choose which missiles to fire at which assets given user-provided Red attack priorities. When Red missile attacks are modeled, TAB-VAM determines the damage to air base infrastructure and capabilities, including the postattack damage to parked aircraft, and then evaluates recovery times and resupply rates. It then calculates how many of the planned sorties each squadron can generate at each air base after each round of Red attacks.

- **Running TAB-VAM and results.** TAB-VAM is typically run for a variety of *attack vectors*, defined in the box to the right. For example, an attack vector can be that ballistic missiles target runways while cruise missiles targets are split evenly between parked aircraft

> **Attack vector:** a set of the adversary's targeting priorities

[8] An analyst could learn to run TAB-ROM without being able to run TAB-VAM if they were working closely with an analyst who can prepare the TAB-VAM input scenario for them. However, in this case, it would still be important for both analysts to be comfortable with both tools' algorithms.

[9] Running TAB-VAM with this information and absent adversary attacks determines the sortie generation for each squadron at each air base over a given period constrained by the air base infrastructure, available resources (e.g., fuel, munitions), and tanker availability. If the sortie generation absent attacks does not match the planned sortie generation, that would indicate shortfalls in available resources and/or infrastructure.

and fuel storage. As one would expect, the choice of attack vector has a large effect on which assets and which air bases suffer the most damage. Furthermore, some mitigations might provide no benefit relative to a given attack vector. (In the example above, mitigations protecting munitions would not provide any benefit because munitions are not part of the attack vector. Thus, it would not be cost-effective to invest in such resources.) Because an adversary's attack priorities are hard to guess, and because it is hugely important in determining the best mitigation strategy, TAB-VAM was designed to make it easy for analysts to run a scenario for multiple attack vectors at once. Figure 2.1. shows a notional set of attack vectors with ballistic and cruise missiles. The allocation of ballistic missiles is shown on the left, and the allocation of cruise missiles is shown on the right. If every combination of ballistic and cruise missiles shown is used, it results in 24 unique attack vectors.

Figure 2.1. Notional Set of Attack Vectors

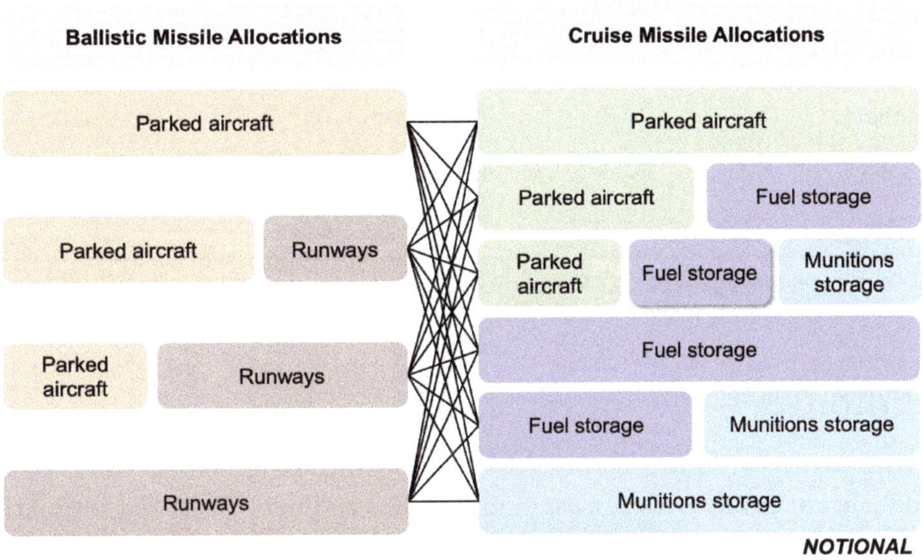

NOTE: The length of each box indicates the proportion of missiles in each allocation.

Although TAB-VAM can be used to answer many analytic questions, it was initially developed to work with TAB-ROM to guide decisions about which investments best mitigate vulnerabilities to a missile attack. Because of this, TAB-VAM provides users with a fair bit of flexibility in defining new mitigations, including active missile defense, runway repair, redundant storage, hardening, decoys, concealment, and medical capabilities. Once defined, mitigations are assigned to locations in a *mitigation strategies table*. Any mitigation that is added to a TAB-VAM mitigation strategies table can be input into TAB-ROM as an investment option. We refer to a set of investment options assigned to specific locations as a *mitigation portfolio*.

Mitigation strategies table: a user input in TAB-VAM that lists the quantity of each mitigation assigned to each location

Mitigation portfolio: the quantity of each mitigation assigned to each location

An example mitigation portfolio is shown in Table 2.1. This table is structured identically to a mitigation strategies table that a user inputs in TAB-VAM. The Location column contains the name of an air base or airfield that the user chose to be part of the portfolio. Headers for subsequent columns specify the names of user-defined mitigation strategies in TAB-VAM. The user can add as many columns as they wish in any order. Prior to running TAB-ROM, the user can provide a default mitigation portfolio that will be incorporated into all the mitigation portfolios that TAB-ROM generates.

Table 2.1. Example Mitigation Portfolio

Location	ADR Kit	Fuel Tank	ECM	Aircraft Shelter	Decoy Fuel Tank	Patriot
Alpha Air Base	1	0	0	5	0	0
Bravo Airfield	1	1	0	0	0	1
Charlie Airfield	1	0	2	6	5	1
Delta Air Base	0	0	0	0	0	0
Echo Airstrip	0	0	2	0	0	0
Foxtrot Airfield	0	0	0	0	0	0
Golf Air Base	1	3	0	0	0	0
Hotel Air Base	0	0	0	0	0	0

NOTE: ADR = airfield damage repair; ECM = earth-covered magazine (used to store aircraft munitions).

The Algorithm

TAB-ROM requires an entire TAB-VAM scenario, including all inputs and the TAB-VAM executable file, a list of attack vectors, a list of mitigations and their costs, and the maximum numbers of each mitigation allowed at each air base (an optional input are placement options for each mitigation, which APT can generate). Given this list of inputs, TAB-ROM will develop and test mitigation portfolios, such as the one shown in Table 2.1. TAB-ROM tests the performance of each mitigation portfolio against each adversary attack via a single run of TAB-VAM. TAB-ROM might run the TAB-VAM scenario upward of a few hundred thousand times as it seeks mitigation portfolios that result in the greatest operational sortie generation at the lowest cost.

Genetic Algorithms

The use of optimization methods is a common approach in determining cost-effective mitigation portfolios. For the problem set we have described thus far, the metric of *effectiveness* (maximizing sortie generation at the lowest cost) is computed by TAB-VAM, which is a complex stochastic simulation model. Because of the inherent nonlinearities found in these types of models, classical optimization approaches—such as linear or convex programming methods—are ill-equipped to tackle the problem. However, heuristic methods, such as GAs, can navigate the solution space—including any inherent nonlinear aspects that space might offer—and converge on optimal solutions in an iterative fashion.[10]

To identify the most cost-effective mitigation portfolio, TAB-ROM uses a GA to address the question of which investments, across a range of bases, will best improve sortie generation, given a specific budget.

GAs share aspects of evolutionary biology, and so does much of the terminology used to describe them. In the language of GAs, a mitigation portfolio is called a *chromosome*. How well a chromosome meets the designed objectives (e.g., high sortie generation, low cost) is called the *fitness* of that chromosome. Each chromosome is composed of *genes*, which in the case of TAB-ROM can be thought of as an investment in a particular mitigation at a particular location (e.g., an investment in three new fuel tanks at a given air base).[11]

A GA will work with multiple chromosomes at once, which together are called a *population*. Continuing the metaphor with evolutionary biology, chromosomes in a population compete with each other and their "parents" to pass their genes on to the next *generation* of chromosomes. GAs determine which chromosomes that pass on their genes through a two-step process. First, GAs rank the chromosomes according to how close they are to the *Pareto frontier*. If the chromosome is on the Pareto frontier, it is assigned a rank of 1. If the chromosome is on the next best frontier, it is assigned a rank of 2 and so on.

Chromosome: a mitigation portfolio

Chromosome fitness: the cost-effectiveness of the chromosome

Genes: the combination of a mitigation, a quantity of that mitigation, and the location

Population: the collection of mitigation portfolios TAB-ROM is working with at one time

Generation: a collection of mitigation portfolios TAB-ROM was previously working with

Pareto frontier: a set of optimal solutions (e.g., most cost-effective mitigation portfolios)

---

[10] Although solutions using algorithms of this nature have not been proven to be optimal, they approach optimal results the longer they are run. We will discuss this iterative solution process later in this section. For more, see David E. Goldberg, *Genetic Algorithms in Search, Optimization, and Machine Learning*, Addison-Wesley Pub. Co., 1989.

[11] This definition of a chromosome for TAB-ROM is over-simplified. The GA does not know about fuel tanks or runway repair kits or even air bases. It simply sees each chromosome as a collection of 1s and 0s that form the binary representation of the number of things being bought at a specific site. Each 1 or 0 in a chromosome is a gene.

Next, the GA needs to identify the parent chromosomes that it will use to create the next generation. To do this, it will first take all the chromosomes from the lowest-ranked frontier. After this, if the number of parents is smaller than the population size, the GA will add all the chromosomes in the next-highest frontier to the parent pool, and so on, until the number of parents is greater or equal to the population size. The GA needs the number of parents to equal the population size, so it will eliminate any excess parents with the smallest crowding distances until the amounts match. After that, GAs eliminate chromosomes of the correct rank, adding them in descending order of crowding distance until there are the same number of chromosomes in the parent pool as the population size. Crowding distance is defined as the distance between a chromosome and its two closest neighbors on the Pareto frontier.[12] Chromosomes on the edges of the frontier are always included, because they have an infinite crowding distance.

Crossover: a random combination of features of two successful chromosomes

Mutation: a random change to an individual feature while the rest of the features remain unchanged

After selecting the set of chromosomes according to the process described above, GAs use *crossover* and *mutation* to generate new chromosomes for the next generation (see the definitions in the box to the right). By combining useful features from different chromosomes, crossover allows a GA to make steady progress toward better mitigation portfolios. However, crossover methods alone can overfocus on local optima, so mutation offers the GA the opportunity to break free and explore broader options for alternative portfolio opportunities that offer improved performance. Unlike crossover in evolutionary biology, TAB-ROM does not swap adjacent sections of its chromosomes. Rather, it pairs off all the chromosomes of the next generation and, based on an input, swaps the corresponding genes based on built-in probabilistic calculations. Mutation is handled a similar way, with genes randomly changing based on the crossover probability, which is hardcoded at 50-percent probability.

For example, one well-performing mitigation portfolio might have placed a runway repair team at Air Base X, while another portfolio has a hardened fighter shelter at Air Base Y. A chromosome in the next generation produced by crossover might include both mitigations, as well as novel features produced from mutation. The GA then continues this process of combining and mutating mitigation portfolios to evolve new, higher-performing ones until a user-defined cutoff is reached. Generally, the cutoff occurs when TAB-ROM has reached a given number of generations without any significant improvement in the fitness score.

Nondominated Sorting Genetic Algorithm

Because TAB-ROM has two simultaneous objectives—maximizing sortie generation while minimizing cost—it is considered a multiobjective optimization. Multiobjective optimizations do not typically have a single global optimal solution, and this is true for TAB-ROM. The question of what the most cost-effective mitigation portfolio is depends, in part, on how much money decisionmakers are willing to spend. It also depends on how good is "good enough" in terms of operational sortie

[12] In this case, the distance is determined using the Manhattan distance measurement, which is a method of calculating distance between points on a grid following the gridlines—that is, calculating every vertical and horizontal step on a grid, not calculating the most direct distance from point to point "as the crow flies."

generation. Because there is no point solution to the problem we pose to TAB-ROM, the tool must seek multiple solutions with each solution having the highest performance possible for its cost.

A subclass within the family of GA methods can address the broader question we raise here: What are the most cost-effective investments to improve sortie generation across a wide range of budgets? This GA subclass is known as a Nondominated Sorting Genetic Algorithm (NSGA). NSGAs can search for cost-effective solutions across a range of possible costs simultaneously.[13] TAB-ROM uses NSGA to evolve a population of mitigation portfolios across a range of costs all at once.

The Pareto Frontier

How close TAB-ROM will get to the Pareto frontier—that is, the optimal solution—depends on the user's input parameters for the algorithm.[14] An example Pareto frontier identified by TAB-ROM is shown in Figure 2.2. The y-axis shows the fitness of the mitigation portfolio.[15] The x-axis shows cost in dollars. Each dot shows the performance of a mitigation portfolio that TAB-ROM evaluated in TAB-VAM. Darker dots show more-recent solutions tested by TAB-ROM, and lighter dots represent older solutions that, in most cases, have been replaced by better solutions. Dots that are in the same generation have the same degree of shading. The red line that connects the dots comprises the Pareto frontier and forms the boundary of the most effective solutions at different investment levels. The newer, darker dots tend to be closest to the Pareto frontier because TAB-ROM finds better solutions over time.

[13] For TAB-ROM, we chose to use NSGA-II. See Deb et al. (2002) for a description of NSGA-II.

[14] To learn more about multiobjective optimization, a useful textbook on the subject is Alessio Ishizaka and Philippe Nemery, *Multi-Criteria Decision Analysis: Methods and Software*, Wiley, 2013.

[15] We discuss this metric and other alternatives available to TAB-ROM users in the next section.

Figure 2.2. Example Pareto Frontier

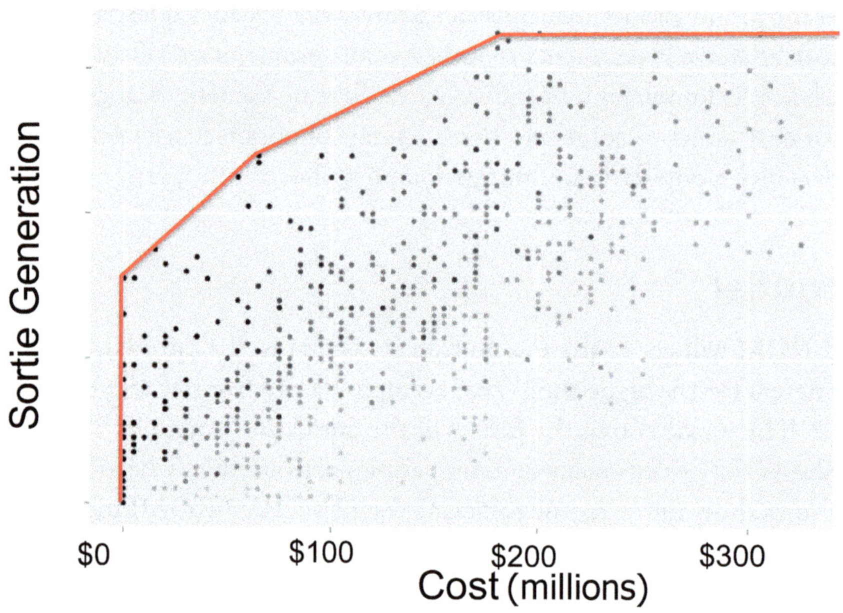

Generally, the Pareto frontier shows significant improvement in performance (here, in terms of sortie generation) with relatively small investments but diminishing marginal returns for increasingly large investments. Therefore, a chart such as the one shown in Figure 2.2 can be a useful decisionmaking aid for determining not just what investments to make but how much money is worth investing. There is little benefit to be gained by investing beyond the knee in the curve.

Objective Functions

Throughout this guide, we have mentioned that TAB-ROM minimizes cost while maximizing effectiveness. In this section, we describe these objectives in more detail.

Minimize Cost

When providing TAB-ROM with a menu of mitigation options, the user must specify as an input the cost of each option. The cost may represent just the procurement cost or an annualized cost, which represents a procurement cost amortized over the expected lifetime of the mitigation combined with yearly storage and maintenance costs.[16]

[16] TAB-ROM can modify the cost of constructing new infrastructure with area cost factors to reflect the fact that building in some locations (such as a remote island in the Pacific) can be more expensive than building in a developed area of an advanced country with a deep civil engineering industry (such as the main island of Honshu in Japan).

Maximize Effectiveness

Analysts can choose how TAB-ROM measures the effectiveness, or fitness, of a mitigation portfolio. When developing TAB-ROM, its creators had to answer two questions: First, which metric produced by TAB-VAM will be used? Second, will multiple TAB-VAM attack vectors be used? And if using multiple attack vectors, how should performance across those attack vectors be incorporated into the performance metric? Some options are discussed below.

Which Metric Produced by TAB-VAM Will Be Used?

TAB-ROM can use only one of TAB-VAM's sortie generation metrics. Prior to running TAB-VAM, the user may choose to include or exclude the performance of active missile defenses, runway availability, the number of aircraft damaged in attacks on parking areas, fuel availability, munitions availability, spares and support equipment availability, or casualties and return-to-duty rates for aircraft maintainers. After selecting which of these capabilities to include in TAB-VAM's model run, analysts can choose one of two sortie generation metrics produced by TAB-VAM to use as the TAB-ROM measure of effectiveness:[17]

- **Location sortie rate:** The percentage of sorties generated, constrained by damage to air bases and recovery timelines, out of the planned number of sorties. Aerial refueling constraints are ignored for this metric even if fuel constraints are included in the TAB-VAM model run. If munitions availability is included in the TAB-VAM model run, aircraft carrying munitions are required to carry a full munitions load to get credit for a sortie flown.
- **Effective sortie rate:** The same metric as the location sortie rate but with aerial refueling constraints taken into consideration (i.e., tanker availability can limit the number of sorties generated) and with partial sortie credit given to aircraft with partial munitions loadouts (e.g., a fighter with half its weapons would get credit for half a sortie, but it still draws a full sortie of fuel).

How Should Performance Across Attack Vectors Be Incorporated into the Performance Metric?

TAB-VAM calculates the above metrics for each attack vector the user specifies. However, TAB-ROM requires a single measure of effectiveness. Therefore, regardless of which sortie generation metric is used, TAB-ROM uses an effectiveness metric that is equal to a weighted average of the sortie generation rate produced by TAB-VAM across the spectrum of attack vectors considered.

Because TAB-VAM is a Monte Carlo model, TAB-ROM runs each portfolio against each attack vector multiple times (typically 25 to 100 times). It then calculates the average performance (e.g., effective sortie rate) across these runs to determine the portfolio's "score" against that attack vector.

To determine the overall fitness of the portfolio against *all* attack vectors, TAB-ROM must balance two considerations: the worst-case performance and overall performance. Under many circumstances, it is reasonable to assume that Red has similar analysis capabilities and will have

[17] TAB-VAM outputs many more metrics that could be used by an optimization model like TAB-ROM; however, their inclusion has not been implemented in TAB-ROM.

identified the attack vector most beneficial to Red—the one that minimizes Blue's objectives (here, sortie generation). From this perspective, TAB-ROM should use the portfolio's worst-case performance as its metric (or value) in its analyses. However, it is also reasonable to assume that some of the assumptions in our analysis might not hold, or that an actual adversary will not choose the same attack vectors our analysis finds to be the most damaging to Blue. Because of this, TAB-ROM weighs the portfolio's overall performance (across all attack vectors) rather than narrowly focusing on a single vector, which might be disfavored by an actual adversary, rest on faulty assumptions, or otherwise be "brittle."

As noted earlier, TAB-ROM calculates the average performance of a portfolio across several TAB-VAM runs for each attack vector to develop a set of performance metrics for a given portfolio. So, instead of considering only the worst-case performance, TAB-ROM averages the performance of the portfolio against each vector but first rounds any results that are more than 10 percentage points greater than the minimum down to that threshold. Given the key role this metric plays in running TAB-ROM, we show an example in Figure 2.3 to help describe how this metric is calculated. The three columns show the performance of three different mitigation portfolios. The y-axis shows the percent sortie generation achieved while under attack—that is, the percentage of planned sorties flown during the TAB-VAM run. The figure shows results from a TAB-VAM scenario using each of these mitigation portfolios against the same set of 24 attack vectors shown in Figure 2.1. Each gray dot represents the percent sortie generation achieved for one attack vector.[18] The red line delineates the worst-performing attack vector for the portfolio; the purple line identifies the sortie generation rate that is 10 percentage points higher than the worst-performing attack vector.

Figure 2.3. Example Chart Showing the Effectiveness of Three Mitigation Portfolios

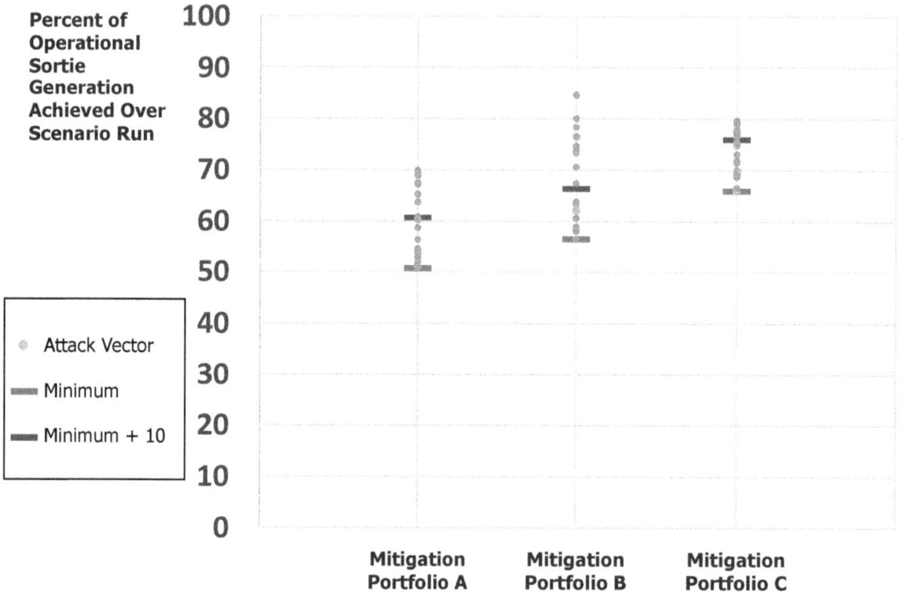

[18] Recall that TAB-VAM is a stochastic simulation. Each gray dot represents the average performance across usually 100 or so Monte Carlo iterations.

Consider Mitigation Portfolio A in Figure 2.3. Across the 24 attack vectors considered, the lowest sortie generation is 50.6 percent, as indicated by the leftmost red line. The purple line marks the value that is 10 percentage points above this minimum sortie generation—specifically 60.6 percent. There are 11 attack vectors against which Mitigation Portfolio A performs better than 60.6 percent sortie generation, which correspond to the 11 gray dots above the purple line. TAB-ROM replaces each of these performance values (sortie generation) with 60.6 percent. It then computes the average performance across all 24 attack vector performance values (where 11 of them have been reduced to a performance of 60.6 percent sortie generation), which results in an overall estimated performance for Mitigation Portfolio A of 57.1 percent. This is the effectiveness metric TAB-ROM uses in its analyses.

Designing our effectiveness metric in this way causes TAB-ROM to focus on improving the performance of the lowest-performing attacks while still allowing the user to recognize some benefit from mitigations that increase the performance of attacks up to 10 percentage points above the minimum.

The choice of effectiveness metric is important, so it is worth a brief discussion of other choices we could have made. We could have instead chosen to average all the attacks, giving them equal weight in TAB-ROM's algorithm.[19] Such an approach would have led TAB-ROM to favor inexpensive mitigations that provide large improvements for the "least damaging" attacks (i.e., attacks that result in the lowest sortie degradation) and spend too little on comparatively expensive mitigations that lead to only modest improvements to the "most damaging" attacks (i.e., attacks that result in greatest sortie degradation). Ultimately, this suggests the potential for "wasted" investment in those inexpensive mitigations, because Red is less likely to opt for attack vectors that do not significantly degrade sortie generation. Blue would still be vulnerable to those most damaging attacks if nothing had been done to mitigate their effects. To illustrate this point, in the example shown in Figure 2.3, there is a large spread in performance. One would expect a smart adversary to avoid attacks against which its opponent is able to achieve at least 80 percent sortie generation (in this example, there are two such attack vectors), especially when there are multiple attacks vectors that can depress sortie generation to 60 percent or less. If that assumption holds, mitigations that further reduce the vulnerabilities of operating locations against the least damaging attacks might not be a wise use of limited funds.

Alternatively, TAB-ROM could have focused on only the most damaging attack by using the minimum sortie rate as its metric of effectiveness in the objective function, thereby ignoring the other attacks. However, the difference in sortie generation between the more damaging attack vectors for a given mitigation portfolio is often small. Focusing on the absolute most damaging attack while ignoring attacks that are *almost* as damaging would prevent TAB-ROM from recognizing a low-cost option to decrease the vulnerability to a slightly less damaging attack. This would not provide the best

---

[19] Assessing the average performance across all attack vectors is known in mathematical parlance as the *L-1 norm*. In comparison, evaluating performance based on the most damaging attack strategy alone is known as the *L-infinity norm*. The approach that TAB-ROM takes lies between these two bookends. The fewer attack strategies considered in developing a mitigation portfolio, the more conservative (and likely expensive) that portfolio becomes. One downside to these more conservative approaches is the risk of uncertainty in how well the chosen portfolio will perform if the adversary chooses not to employ that singular attack strategy. For more information on varying norms in objective functions and the risks in their application, refer to Stephen Boyd and Lieven Vandenberghe, *Convex Optimization*, Cambridge University Press, 2004.

information to decisionmakers looking to make real-world investments to mitigate attacks in a possible real conflict.

TAB-VAM and TAB-ROM rely on many assumptions about how a future conflict would play out, yet we can never know if a real adversary would make the same calculation that we have. Considering a larger number of possible attacks in TAB-ROM's effectiveness metric helps guard against this uncertainty.

Constraints

TAB-ROM allows analysts to specify constraints for the GA to follow when developing candidate mitigation portfolios. Analysts can use such constraints to prevent unrealistic solutions—for example, adding an absurd numbers of munition storage shelters to a small airfield. They can also be used to limit the number of possible solutions, which will allow TAB-ROM to converge on the optimal solution sooner.

Mitigation Increment Size

When analysts provide the menu of investment options to TAB-ROM, they must specify the increment size by which TAB-ROM increases or decreases the quantity of the investment at an operating location. For example, the user can require aircraft shelters to be purchased one at a time while fuel bladders must be purchased in sets of four. Larger increment sizes limit the flexibility of the GA to find the best solution, but smaller increment sizes, especially for low-cost mitigation options, typically result in longer run times because the GA will need to search a larger space of candidate solutions. Analysts specify the mitigation increment size in the Genes file, which we describe in Appendix A.

Mitigation Upper Bound by Location

As discussed earlier, users can provide upper limits on the quantity of a given mitigation that can be purchased for an operating location (see Table 2.1 and the associated discussion). These upper bounds define the highest-cost mitigation portfolio TAB-ROM can consider. Analysts specify the mitigation upper bound by location in the Genes file, which we describe in Appendix A.

Upper bounds are another means by which the analyst can reduce run time by limiting the search space of the GA. It is also a way to prevent unreasonable solutions at a given scenario. One real-world constraint might be the amount of space available for additional infrastructure and other mitigations to be placed at an operating location. Analysts can use a tool, such as APT, to estimate the space available for each mitigation type and then use that tool's output as an upper bound in TAB-ROM.

Space Constraints

TAB-ROM also lets users include mitigation placement options, typically generated by APT. These options are part of the TAB-VAM inputs and are included in the full TAB-VAM documentation, but they can play an expanded role when used with TAB-ROM.

Providing mitigation placement data from APT allows TAB-VAM to assign mitigations to a specific location in the simulation (specified by x, y coordinates relative to a center point on the airfield). When the GA generates a chromosome that includes different types of assets at the same base for which placement data is provided, TAB-ROM attempts to find sets of locations for these assets so that they do not interfere with other assets, such as by determining the required safety distances between assets (especially with fuel and munitions). If it is not possible to fit all of the assets (of all types) identified by the GA, TAB-ROM will proportionally reduce the number of assets of each type to identify a physically feasible solution with the same asset mix. The chromosome is then modified to match these reduced numbers, and the GA continues. By forcing TAB-ROM to respect the physical constraints of the air base, we enable it to explore the trade-offs inherent in those constraints (i.e., whether to fill available space with one asset or another, or a combination).

In addition to limiting consideration to only those combinations of investments that are physically feasible, placement of mitigation options allows for collateral damage calculations to be performed within the TAB-VAM runs of the analysis, thereby enhancing the results' fidelity.

Events of a TAB-ROM Run

In TAB-ROM, the GA begins with a quasi-random selection of candidate chromosomes (i.e., mitigation portfolios).[20] These chromosomes include (1) an empty chromosome that includes only the default mix of assets at each location, (2) a "max" chromosome, which includes the maximum number of each asset that can be purchased for each location, and (3) a chromosome for each asset-location pair that includes the pair's upper bound and the default mix for all other pairs. If these chromosomes number less than the user-specified population size (e.g., the maximum number of chromosomes per generation), crossover chromosomes are generated by randomly selecting genes until the population is equal to the population size.

Evaluating the Initial Population

At each step in the GA, each member of the population is evaluated by running TAB-VAM and computing the specified "goal" metric, such as the effective sortie rate.[21] Once TAB-ROM has scored each candidate mitigation portfolio in the initial population, it identifies the *efficient frontier* (i.e., the

---

[20] Instead of starting with a random selection of candidate mitigation portfolios, analysts can seed TAB-ROM with the population of mitigation portfolios. Typically, the population information will be taken from a previous TAB-ROM run. This enables TAB-ROM to "pick up where it left off" if it was halted during the process of searching for the Pareto frontier.

[21] As noted above, a weighted average of performance across attack vectors is used. The weighted average discounts the value of any attack vector performance that is greater than 10 percentage points above the minimum. Red's attack vector is a key input to TAB-VAM, and uncertainty about Red's strategy is a defining assumption of these models. This uncertainty is even greater in the context of TAB-ROM because different mitigation strategies by Blue dramatically alter what will be most effective for Red.

mitigation portfolios that have the better performance than any other mitigation portfolio of equal or lower cost).

Creating Further Generations

After evaluating the initial generation, TAB-ROM creates an equal number of "children" chromosomes to populate the next generation. These new chromosomes and chromosomes that were located at the previous Pareto frontier then undergo the GA's selection process (described in more detail in Chapter 3). The remaining chromosomes then become pairs of new parent chromosomes. The number of these new parents equals the population size the user had preconfigured. These pairs are exclusive—that is, each parent is only in one pair. Crossover is performed on each of the pairs to generate two new chromosomes in which the new chromosomes are equivalent to one of their parents with some number of genes swapped in from the other parent (dependent on the user-defined crossover rate). Random mutations are also inserted at a user-configured rate.

Each of these child chromosomes is then evaluated against the mitigation portfolios and added to the total population so that TAB-ROM can recompute the Pareto frontier. TAB-ROM selects a new population from this frontier, and it will also select chromosomes of lower rank if there are insufficient chromosomes on the frontier (as discussed, TAB-ROM uses crowding distance as a selection metric). New pairs of parents are identified from this updated frontier and the process continues.

Ending the Model Run

Throughout the run, TAB-ROM checks to see how much improvement took place between the current generation and the previous generation. When it sees that there is approximately no improvement for a user-specified number of generations (which the user sets with the generation stall limit input variable), the run will end.

The user can also check in on TAB-ROM's progress throughout the run and choose to end it early. For example, an analyst might decide to end a TAB-ROM run early if only a small number of chromosomes in each generation improve the Pareto frontier and that improvement is small.

TAB-ROM outputs the mitigation portfolios (i.e., the type and number of mitigations purchased for each operating location) that form the Pareto frontier along with the cost and sortie generation for each. It also outputs the mitigations that were purchased for the chosen operating locations.[22] These outputs, and others, are described in Chapter 4.

[22] The default mix of assets at each location is not included in the mitigation portfolios that TAB-VAM outputs. If the analyst wants to run a mitigation portfolio in TAB-VAM, the default mitigations will need to be added.

Chapter 3

Inputs

This chapter describes each input for TAB-ROM. Appendix A offers more detail on each of the algorithms, and the chapter following this one describe these algorithms as they are executed in TAB-ROM.

TAB-VAM

All the inputs required to run a TAB-VAM scenario, including the TAB-VAM executable file, are required to run TAB-ROM. For ready reference, we have included a brief description of how TAB-VAM's algorithms and techniques interact with TAB-ROM can be found in Chapter 2.

TAB-ROM Inputs

TAB-ROM reads all inputs from the *inputs* directory, which the user specifies when the model is run. By manipulating the files in this directory, the user sets the parameters and assumptions for model runs.

Table 3.1 describes the seven main TAB-ROM input files. The first column contains the name of the input. The example path indicates the typical location and name of the file. The description contains a brief overview of the file's contents.

The top level of the TAB-ROM *inputs* directory contains the files required to run TAB-ROM. Several file formats are used. One file is in extensible markup language (XML) and has the file extension *.xml*. Two files are in JavaScript Object Notation (JSON). TAB-ROM uses a syntax expansion to JSON called Human JSON (HJSON), and so these files end with the file extension *.hjson*. XML and JSON files can be viewed and edited with any text editor, several of which are discussed in Appendix C. These editors are readily available online at no cost. The inputs directory also contains data tables stored as comma-separated values (CSV) files. CSV files end with the file extension *.csv* and are most easily edited with a spreadsheet program, such as Microsoft Excel. Note that other files and directories may be stored within the TAB-ROM inputs directory but TAB-ROM will ignore any file or directory that it is not expecting.

Table 3.1. TAB-ROM Input Files

Short Description	Input	Example Path and File Name	Description
Model parameters	Parameters	../inputs/Parameters.xml	The parameters for the GA and the file names and paths for all other TAB-ROM and TAB-VAM input files to be used.
Optimization parameters	Objectives	../inputs/Objectives.hjson	Indicates which objective functions TAB-VAM should use. Options include cost in combination with one sortie generation metric as calculated by TAB-VAM.
Red attacks	Allocations	../inputs/tblAllocations.csv	The list of attack vectors.
Menu of investment options	Genes	../inputs/Genes.hjson	The list of mitigations along with their costs and the increments in which they can be purchased.
	Cost Multipliers	../inputs/CostMultipliers.hjson	Defines the location-specific cost multipliers that should be applied to investments.
Constraints	Defaults	../inputs/tblDefaults.csv	The list of mitigations at each location prior to any additional investments. These investments will be a part of every mitigation portfolio that TAB-ROM creates.
	Upper Bounds	../inputs/tblUpperBounds.csv	The maximum number of each mitigation that can be purchased for each location. These upper bounds are not necessarily connected to the location available as defined by APT.

Optimization Parameters

TAB-ROM requires that the user choose a set of metrics that will guide the selection of elements within the mitigation portfolio. The metrics are typically cost and sortie rate, but they can also be the number of people died of wounds, number of people returned to duty, and individual location sortie rate. The objective function plays a central role in the optimization process. For example, in most CODE analyses, the objective is to maximize all sorties theaterwide across a range of attack vectors. This means that the objective function does not differentiate between aircraft types, so a special operations forces sortie is given the same weight as an intelligence, surveillance, and reconnaissance sortie, and so on. A bomber's long-range strike mission is given the same importance as a close-in fighter aircraft's defensive counter air sortie.

It is also important to note that TAB-ROM, like TAB-VAM, does not assess the combat *effectiveness* of sorties but instead just computes the sortie generation *capability* provided by investments in damage prevention and mitigation options.

Red Attacks

An adversary's sortie degradation can vary significantly between different targeting plans, and as with TAB-VAM, TAB-ROM allows the user to specify a set of adversary attack vectors.[23] The typical objective function in a CODE analysis is to maximize sortie generation across a set of user-defined attack vectors. For example, analysts might seek mitigation portfolios that are resilient against attacks on runways, parking, and fuel systems. Diversifying the portfolio can help protect the base as a whole and mitigate the risk of damage to its resources. Although such an approach might not be the most effective solution against any individual attack vector, the portfolio can be designed to generate the highest number of average sorties when played against a range of attack vectors in the aggregate.[24]

Menu of Investment Options

TAB-ROM requires the user to provide a list of specific investment options along with the cost of each option. Investment options may include any mitigation recognized by TAB-VAM's Mitigations table. Other mitigation strategies, such as modifying the air base posture, employing force maneuver CONOPS, adjusting the flow of aircraft into the theater, or modifying other parameters, cannot be included as an investment option for TAB-ROM.

Location-specific cost multipliers may also be specified to adjust for regional cost differences in investments, particularly for infrastructure. For example, building new infrastructure in Germany might not cost the same as in Italy. Analysts can group locations however they like (e.g., by country, by region).

Constraints

Constraints to TAB-ROM's mitigation portfolios come in two forms: (1) default mitigations cannot be removed or relocated, and (2) there are limits, or upper bounds, on the number of resources of a given mitigation type that can be purchased for a given location (based in part on the physical constraints of the location).

---

[23] See Figure 2.1 for an example set of adversary attack vectors that can be used to automatically generate attack plans when using TAB-VAM. The user also has the option of specifying a single attack plan.

[24] This development of mitigation portfolios that are robust against uncertainty about an adversary's attack strategy is consistent with risk-management strategies for financial portfolios. These risk-versus-return approaches are well documented in finance literature (as in Harry M. Markowitz, *Portfolio Selection: Efficient Diversification of Investments*, Yale University Press, 1971).

Default Mitigations

Users might wish to specify existing mitigations that TAB-ROM might not remove or relocate. For example, the force laydown used for the analysis might include active missile defense systems assigned to protect specific air bases. Similarly, a site survey might determine that existing fuel storage tanks at an older airfield have degraded and require replacement. This might indicate a minimum number of new tanks that must be constructed at the location. The user can incorporate any such known infrastructure lower bounds into TAB-ROM's procurement options.

Upper Bounds

The user can specify any constraints on mitigation portfolios that TAB-ROM explores. The user may limit the total quantity of a mitigation purchased theaterwide or the number of assets procured at a specified location.

A key limiting factor for theaterwide investments is the availability of USAF manpower and equipment. For instance, if the analyst learns there is a limit in manpower or equipment for Fuels Operational Readiness Capability Equipment kits, they can limit the total number of kits that TAB-ROM can procure.

Availability of physical space at a base is another significant category of constraints. After reviewing site surveys and beddown planning processes, the user might discover that there are limited opportunities for constructing new infrastructure at a base. For example, small bases and locations near dense urban regions might lack sufficient footprints to accommodate new, large structures, such as static fuel tanks or hardened aircraft shelters. Thus, the user might want, or need, to constrain the number of these assets that can be placed at a given operating location.

Upper bounds also allow the user to prohibit TAB-ROM from investing in certain mitigations at specific bases. For example, the user can choose to tailor the placement of earth-covered munitions storage only at bases within the inner missile threat ring, or the user can restrict procurement of hardened fighter shelters to only those bases hosting fifth-generation fighters.

TAB-ROM is flexible enough to accommodate most user-defined constraints. For example, if the U.S. Marine Corps is investing in advanced airfield damage repair capabilities different from those of the USAF, the Marine Corps variant can be prohibited at the USAF operating locations by setting an upper bound of zero (and vice versa). Similarly, in this way, ally investment strategies can be developed in conjunction with U.S. investments by restricting which locations can buy ally assets and which can buy U.S. assets.

Using APT to Quantify Space Constraints

As suggested above, availability of space, or lack thereof, can limit the number of additional assets—such as fuel infrastructure, munitions storage, runways, or parking aprons—that can be accommodated. APT can be used to determine an upper bound on mitigation portfolios based on spatial constraints.

APT requires two key pieces of information: (1) the necessary dimensions for installing and operating investment assets and (2) a geospatial boundary within an air base that represents an area that does not contain any infrastructure or assets. The user inputs the number of assets to be placed and their dimensions. Assets are packed into unoccupied spaces, one by one, until there are no more assets to be placed, or there is no more space available. The placement rules are defined by two parameters: the grid spacing and the minimum separation distance. The grid spacing defines the spatial resolution of the usable area. Each point in the grid represents a potential space to be occupied by an asset. The minimum separation distance defines how close assets can be placed to one another. The result of running APT on each base provides an additional spatial constraint on the number of investment assets that the base can support.

Chapter 4

Outputs and Interpretation

In this chapter, we describe TAB-ROM outputs that users can expect at the end of a run. We also offer guidance for analysts to consider while interpreting the outputs and trends to be aware of when conducting future TAB-ROM analysis.

TAB-ROM Output Tables

On completion, TAB-ROM generates six output tables that are returned in CSV files, as described in Table 4.1. Appendix B provides a more detailed discussion of the tables.

Table 4.1. Output Tables

Category	Output Table Name	Table Description
Mitigation portfolio contents	Chromosome-Gene	Reports the list of genes in each chromosome.
	Genes	Describes each gene, which is the quantity of a mitigation at a location. For example, "Five aircraft shelters at Andersen Air Force Base" can be a gene.
Mitigation portfolio performance	Chromosome	Reports the cost and effectiveness of each chromosome (i.e., mitigation portfolio).
	Chromosome ESRS	Reports the effective sortie rate achieved for each attack vector for each chromosome.
GA data	Chromosome Generation	Contains information on the GA-related aspects of each chromosome. The table columns include each generation's chromosome ID, rank, and crowding distance.

NOTE: ESRS = effective sortie rate(s).

Mitigation Portfolio Contents

The mitigation portfolios are described in two output tables: the Chromosome-Gene output table and the Gene output table. The Chromosome-Gene output table contains the list of genes that make up each chromosome. The Gene output table describes each gene. Recall that a gene is the combination of a location, a mitigation, and the quantity of that mitigation at that location. The same gene will likely appear in many chromosomes. When postprocessing the outputs, the user can combine these two files to see the quantity of each mitigation assigned to each location for each mitigation portfolio.

Mitigation Portfolio Performance

The Chromosome Output table contains a list of all mitigation portfolios run by TAB-VAM and records the cost and effectiveness of each mitigation portfolio.

The effectiveness metrics reported in the Chromosome table are those used by the GA, which represent an average across the full range of attack vectors. For deeper dives or targeted analysis, the analyst might require more granular data at the attack vector level. The Chromosome ESRS output table records the effective sortie rate performance of each mitigation portfolio against each of the attack vectors. The effective sortie rate is calculated over several steps. First, if the sortie rate for an attack vector is over 10 percentage points higher than the minimum sortie rate in the set of allocations, TAB-ROM will set it to be 10 percentage points higher than the minimum sortie rate. Then, TAB-ROM will take the average of all the sortie rates.

GA Data

The Chromosome Generation Output table contains information about GA-related aspects of each chromosome. It identifies the generation to which each chromosome belongs; the chromosome's rank (i.e., how the chromosome's fitness score compares with others in the generation); and the chromosome's crowding distance, which describes the distance between the chromosome's fitness score and its nearest neighbors. Crowding distance is a useful metric for GAs. Points that are crowded (i.e., points that are all bunched up at one place on the Pareto frontier) are not as likely to get passed on in the next generation because they are not fully spread out along the Pareto frontier (see Chapter 2 for an explanation of how new generations are created).[25]

Interpreting Outputs

Once analysts have generated outputs from TAB-ROM, they confront the critical step of interpreting them. Although the full research design of an analysis is outside the scope of this guide for using the TAB-ROM tool, we offer some lessons we have drawn from using TAB-ROM over the past decade to conduct cost-effectiveness analyses.

Analysts should bear in mind that because TAB-ROM uses a GA, we cannot guarantee that the tool has found *the* optimal mitigation portfolio at each budget level. But if TAB-ROM has been run until solutions *converge* with the Pareto frontier, analysts can be confident that the tool has found cost-effective portfolios at all the budget levels along the frontier. Combined with the greater challenge of ensuring a proper match between substantive policy issues the government faces and the formal

---

[25] An example with good pictures and explanations can be found in Wei Gong, Qingyun Duan, Jianduo Li, Chen Wang, Zhenhua Di, Aizhong Ye, Chiyuan Miao, and Yongjiu Dai, "Multiobjective Adaptive Surrogate Modeling-Based Optimization for Parameter Estimation of Large, Complex Geophysical Models," *Water Resources Research*, Vol. 52, No. 3, March 2016.

problem TAB-ROM addresses, we recommend that analysts focus on the trends in what TAB-ROM recommends buying along the frontier rather than narrowly fixating on the specific details of any one portfolio.

Another dynamic that analysts should remember when interpreting TAB-ROM results is a phenomenon we have called *gap filling*. If the investment options have wide variance in their costs (e.g., some options cost tens of thousands of dollars yet others cost tens of millions of dollars), then TAB-ROM results can include widely fluctuating quantities of the low-cost investment options as the budget changes along the frontier. Because inexpensive options generally have smaller effects than more expensive options, TAB-ROM will sometimes decrease the quantity of inexpensive options bought to add a more expensive option. In these cases, analysts should explore which investment options are the more significant force driving results. Often, the more expensive options will create most of the sortie generation gains, while the large numbers of inexpensive options yield increases in sortie generation on the order of merely fractions of a percentage point. Because TAB-ROM tries to maximize sortie generation at every budget level, it will add portfolios that increase sortie generation by fractions of a percentage point, even when the real-world implications of that increase are likely within the uncertainty bounds of the underlying inputs for the TAB-VAM scenario. Therefore, we encourage analysts to identify those options that have the largest effect on sortie generation and focus recommendations on those results instead of focusing on the large and fluctuating quantities of inexpensive options that have marginal effects.[26]

Additionally, we encourage analysts to look at trends across the Pareto frontier by considering the following:

- **At what points on the frontier does a given asset type appear and stay in portfolios along the frontier for good?** Identifying the investment levels after which an asset remains present in more expensive portfolios along the Pareto frontier can indicate the overall investment level above which it will always be cost-effective to include the asset in the portfolio of investments. For example, if TAB-ROM shows that a frontier portfolio at the $300 million budget level "bought" 427 fuel bladders across the theater, but similarly priced portfolios bought substantially different quantities of fuel bladders (e.g., a $298 million portfolio bought only 300 fuel bladders), we would not take 427 or 300 to be our final answer. However, if TAB-ROM always bought at least 400 fuel bladders once its budget exceeded $350 million, then we would feel confident concluding that, at budget levels above $350 million, there was a strong cost-effectiveness case for procuring at least 400 more fuel bladders.
- **Which asset types are purchased first (in the lower-cost portion of the frontier)?** Unsurprisingly, when TAB-ROM is offered a variety of different investment options that all address a similar vulnerability, it buys the least expensive asset types first. For example, TAB-ROM will purchase the least expensive active defense options before buying any other type of active defense system. When choosing between mitigations that address different vulnerabilities, sometimes TAB-ROM will find a few investments that dominate spending on

[26] One way to explore which options have the largest sortie generation effect is to take a set of Pareto frontier–optimal portfolios identified by TAB-ROM and run a sensitivity analysis on them in TAB-VAM. In this sensitivity analysis, an analyst would conduct a series of TAB-VAM runs and remove elements of the TAB-ROM portfolio individually to see what the marginal effect of each element of the portfolio was.

the lower-cost part of the fronter (i.e., the portion of the frontier that exhibits the quickest increase in sortie generation). One way to recognize this is by examining how much of each portfolio's budget is spent on each asset type. If TAB-ROM spends over 50 percent of the total budget of all portfolios on the lower-cost portion of the frontier on a single asset type, then that is compelling evidence that the asset type is highly cost-effective.

Appendix A

Input Details

In this appendix, we describe in detail each input file we discussed in Chapter 3.

Parameters File

The Parameters file used by TAB-ROM is typically named *Parameters.xml*, but its name and location are specified in the shell script used to call TAB-ROM, so it could technically be given any name. (See Appendix C for instructions on how to modify the shell script.) This is the file that TAB-ROM will look for at the start of each run. Several parameters specified in the Parameters file are file paths to the input and output files that TAB-ROM will need to be able to run. For simplicity and portability, we recommend specifying file paths relative to the default *inputs* or *outputs* directory as indicated in the description below. Using absolute file paths will make it difficult to make changes to input files or reuse the TAB-ROM input and output directories for new scenarios. Table A.1 describes the parameters in the Parameters file, and Figure A.1 depicts a notional file.

Table A.1. Parameters File Description

Category	Parameter	Data Type	Format	Description
Genetic algorithm parameters	Population size	Integer	intPopulationSize	The number of individual mitigation portfolios that will be run in each TAB-ROM generation (we recommend at least 25).
	Max steps	Integer	intMaxSteps	The max number of generations before the run will end.
	Threads	Integer	intThreads	The number of threads. A *thread* is a term referring to a part of an overall process that is semi-independently managed by the user's computer's scheduler. In this case, this is each individual instance of TAB-VAM that gets run at a given time.
	Timeout (ms)	Integer	intTimeoutMS	The longest time, in milliseconds, that any individual thread (TAB-VAM run) will be allowed to run.

Category	Parameter	Data Type	Format	Description
	Verbosity	Integer	intVerbosity	An input that determines if output tables will be printed or not. If the integer provided is a zero or greater, output tables will be printed. If the integer provided is less than zero (i.e., a negative number), no output tables will be printed.
	Random seed	Integer	RandomSeed	The initial seed for random number generation.
NSGA parameters	Stall generation limit	Integer	intStallGenerationLimit	The number of generations over which improvement has to be stalled before TAB-ROM halts.
	Stall tolerance	Double	dblStallTolerance	The level of improvement, stated as a percentage, that constitutes stalling for the stall generation limit.
	Probability of mutation	Double	dblProbabilityOfMutation	The probability of each gene instance mutating to another gene instance in the same gene. That is, the probability that the quantity of a specific mitigation at a specific location will randomly change in the next generation.
TAB-VAM inputs	TAB-VAM input directory	String	strVAMInputDir	The location of the directory, relative to the directory from which the run command is executed, that contains the TAB-VAM input files required to run the scenario used to assess the performance of the mitigation portfolios.
	TAB-VAM parameter file	String	strVAMParametersFile	The name of the XML file in the top level of the TAB-VAM input directory required to run the scenario used to assess the performance of the mitigation portfolios.
	TAB-VAM resources directory	String	strVAMResourcesDir	The location of the directory, relative to the directory from which the run command is executed, that contains additional files that TAB-VAM requires to run.
TAB-ROM inputs	Cost multiplier file	String	strCostMultiplierFile	The relative file path to the Cost Multiplier file that TAB-ROM needs to run from the directory where the command was run.
	Genes file	String	strGeneDefaults	The relative file path to the Genes file that TAB-ROM needs to run from the directory where the command was run.
	Defaults table	String	strDefaultTable	The relative file path to the Defaults table that TAB-ROM needs to run from the directory where the command was run.

Category	Parameter	Data Type	Format	Description
	UpperBound table	String	strUpperBoundTable	The relative file path to the UpperBound table that TAB-ROM needs to run from the directory where the command was run.
	Allocations table	String	strALlocationsTable	The relative file path to the Allocations table that TAB-ROM needs to run from the directory where the command was run.
	Objectives file	String	strObjectiveFile	The relative file path to the Objectives file that TAB-ROM needs to run from the directory where the command was run.

Figure A.1. Example Parameters File

```
<xml>
  <!-- GA parameters -->
  <intPopulationSize>2</intPopulationSize>
  <intMaxSteps>200</intMaxSteps>
  <intThreads>6</intThreads>
  <intTimeoutMS>300000</intTimeoutMS>
  <intVerbosity>0</intVerbosity>
  <randomSeed>0</randomSeed>

  <!-- NSGA parameters -->
  <intStallGenerationLimit>40</intStallGenerationLimit>
  <dblStallTolerance>0.01</dblStallTolerance>
  <dblProbabilityOfMutation>0.01</dblProbabilityOfMutation>

  <!-- ROM inputs -->
  <strCostMultiplierFile>CostMultipliers.hjson</strCostMultiplierFile>
  <strGeneFile>tblGenes.csv</strGeneFile>
  <strDefaultsTable>tblDefaults.csv</strDefaultsTable>
  <strUpperBoundTable>tblUpperBounds.csv</strUpperBoundTable>
  <strAllocationsTable>tblAllocations.csv</strAllocationsTable>
  <strObjectiveFile>Objectives.hjson</strObjectiveFile>

  <!-- VAM inputs -->
  <strVAMInputDir>depends/lvm/inputs</strVAMInputDir>
  <strVAMParametersFile>Parameters.xml</strVAMParametersFile>
  <strVAMResourcesDir>resources</strVAMResourcesDir>
</xml>
```

Objectives File

The Objectives file used by TAB-ROM is typically named *Objectives.hjson*. The name and location of the Objectives file is specified in the Parameters file. This file, described in Table A.2, contains Boolean-type information to indicate whether cost should be considered within the objective function and which effectiveness metric should be used. Although the Cost indicator should be set to "true" (to ensure TAB-ROM aims to minimize cost), the user could opt to omit Cost from the objective function and allow TAB-ROM to purchase everything to determine the maximum sortie generation possible. Only one effectiveness metric should be used. (The user could technically include both sortie rates as objectives and obtain a TAB-ROM manifold instead of a curve. However, this would likely result in extremely long run times, and with both objectives being close variations on sortie rate, the end results might not be particularly interesting.) Table A.2 describes the Objectives file, and Figure A.2 shows a notional example of such a file.

Table A.2. Objectives File Description

Category	Parameter	Data Type	Format	Description
Objectives	Cost	Boolean	*true*	true will use cost in the objective function
			false	false will not consider cost in the objective function
	EffectiveSortieRate	Boolean	*true*	true will use TAB-VAM's EffectiveSortieRate in the objective function
			false	false will not consider TAB-VAM's EffectiveSortieRate in the objective function
	LocationSortieRate	Boolean	*true*	true will use TAB-VAM's LocationSortieRate in the objective function
			false	false will not consider TAB-VAM's LocationSortieRate in the objective function
	DiedOfWounds	Boolean	*true*	true will use TAB-VAM's DiedOfWounds in the objective function
			false	false will not consider TAB-VAM's Died of Wounds in the objective function
	ReturnToDuty	Boolean	*true*	true will use TAB-VAM's ReturnToDuty in the objective function
			false	false will not consider TAB-VAM's ReturnToDuty in the objective function

Figure A.2. Example Objectives File

```
{
 objectives: [
  {
   name: Cost
   active: true
  }
  {
   name: EffectiveSortieRate
   active: false
  }
  {
   name: LocationSortieRate
   active: true
  }
 ]
}
```

Allocations Table

The Allocations table used by TAB-ROM is typically named *tblAllocations.csv*. The name and location of the Allocations table file is specified in TAB-ROM's Parameters file.

The Allocations table lists the attack vectors that TAB-VAM will use to evaluate each mitigation portfolio. The user does this by pointing to TAB-VAM's Automatic Allocation table that describes the attack vector to be used and giving that vector a name.

Users can, and typically do, build a list of attack vectors in the Allocations table. Because each attack vector is run separately by TAB-VAM, the number of attack vectors affects how many TAB-VAM runs TAB-ROM will need to make, and therefore how long it will take for TAB-ROM to converge on an answer. For example, doubling the number of attack vectors doubles the number of times TAB-VAM will need to be run to test each mitigation portfolio.

Table A.3 describes the Allocations table.

Table A.3. Descriptions of Allocations Table Columns

Table Name	Allocations		
Example Path	tblAllocations.csv		
Table Description	The Allocations table contains the list of Red missile allocations to be run by TAB-VAM. Each row includes a user-selected Allocation Name and the path to the file defining the Red attack associated with the allocation.		
Column Name	**Data Type**	**Format**	**Column Description**
Name	String	Can be built of letters, numbers, or dashes. Other special characters, such as commas, underscores, and spaces, should not be used.	The user-defined allocation name. This name will be used to identify outputs associated with the allocation in the output files.
File name	String	File path	The path to the Phase Allocation file relative to the TAB-VAM inputs directory. This will be used instead of the Phase Allocation file assigned in TAB-VAM's Campaign file.

Genes File

The Genes file is typically named *Genes.hjson*. The name and location of the Genes file is specified in the Parameters file described previously. This file lists the mitigations available to TAB-ROM, specified at the highest level as *assets* or *choices*, along with parameters associated with those mitigations. Mitigations for which there is only one type of *option* (e.g., fuel tanks) are specified as assets, whereas mitigations for which there are multiple, mutually exclusive options from which to choose (e.g., runway repair capabilities come in multiple forms and sizes, but only one can be assigned to a given location) are specified as choices. Associated parameters are specified for the individual options for that capability. (This distinction exists only in TAB-ROM and is not reflected in TAB-VAM; some mitigations, such as missile defense, can be included as assets or grouped into choices.) Table A.4 describes the parameters in the Genes file. Figure A.3 contains an example Genes file.

Table A.4. Genes File Description

Parameter	Data Type	Format	Description
Name	String	The name of a user-defined asset or capability defined in one of the following TAB-VAM input tables: Aeromedical Evacuation, Aircraft Decoys, Escort Ships, Fuel Storage Defaults, Medical Facilities, Missile Defenses, Pipeline Repair, Runway Repair, Shelter Defaults, and Supply Ships	The name of a mitigation that meets one or both of the following criteria: 1. It is listed in the Defaults table. 2. TAB-ROM should consider it as a part of a mitigation portfolio.

Parameter	Data Type	Format	Description
Cost	Double	Zero or any positive number	The cost of each mitigation.
Buy-increment	Integer	Any positive integer	The increment at which the mitigation can be purchased.
Number-free	Integer	Zero or any positive integer	The number of this mitigation (asset or option) that can be included in a portfolio (across all locations) cost-free. That is, the cost of this mitigation is not included in the total portfolio cost until this number is exceeded. The number defaults to zero if not specified. This parameter is for use either when assets have already been purchased but not placed or when existing assets are mobile and could be placed in different locations.

Figure A.3. Example Genes File

```
{
  assets: [
    {
      name: BLADDER
      cost: 0.2
      buy-increment: 10
      number-free: 100
    }
  ]
  choices: [
    {
      name: Runway Repair
      options: [
        {
          name: CIVE
          cost: 0
        }
        {
          name: Small CRATR
          cost: 10
        }
        {
          name: Medium CRATR
          cost: 20
        }
     }
]
```

Cost Multipliers File

The Cost Multipliers file used by TAB-ROM is typically named *costmultipliers.hjson*. The name and location of the Cost Multipliers file is specified in TAB-ROM's Parameters file.

The Cost Multipliers file is used to identify regional cost adjustments for mitigations and contains three distinct sections: asset groups, location groups, and multipliers. Nonrecurring infrastructure costs give the user the option of providing an additional nonrecurring cost multiplier for infrastructure mitigations that are distinct for infrastructure projects.

The asset-groups portion of the file identifies sets of assets that may be subject to the same location-based adjustment. The user can specify an arbitrary number of different asset groups that are characterized by the group name and the list of assets in the group. Each asset can only be assigned to a single group. Every asset must be assigned to an asset group.

The location-groups portion of the file can similarly contain an arbitrary number of different location groups, each of which is defined by a group name and the list of locations in the group. Each location can only be in a single location group. Locations not assigned to a location group are assumed to have multipliers of 1.

The multipliers portion of the file contains a complete set of pairings between the asset and locations groups, which identifies a regional cost multiplier for each set, as shown in Figure A.4. Within TAB-ROM, this multiplier is applied to all assets of the asset group that are placed in any location within the location group. Table A.5 describes the Cost Multiplier file.

Table A.5. Cost Multiplier File Description

Tag	Name	Data Type	Format	Description
Asset-groups	Name	String	A string name of a user-defined group of assets	The user-defined name for the asset group
	Assets	String	The name of a user-defined asset or capability defined in one of the following TAB-VAM input tables: Aeromedical Evacuation, Aircraft Decoys, Escort Ships, Fuel Storage Defaults, Medical Facilities, Missile Defenses, Pipeline Repair, Runway Repair, Shelter Defaults, and Supply Ships	The list of mitigations in the group
Location-groups	Name	String	A string name of a user-defined group of locations	The user-defined name for the location group
	Location	String	The name of a user-defined location defined in TAB-VAM's Locations table	The list of locations in the group
Multipliers	Asset-group	String	The name of a user-defined asset group in this file	The name of the asset group
	Location-group	String	The name of a user-defined location group in this file	The name of the location group
	Multiplier	Double	Any positive number	The "cost factor." The mitigation's cost will be multiplied by this number to get the location-specific cost for the mitigation for any locations in the location group.

Figure A.4. Example Cost Multipliers File

```
{
 asset-groups: [
  {
   name: shelters
   assets: [
    Small Shelter
    Large Shelter
    Expeditionary Small Shelter
    Expeditionary Large Shelter
   ]
  }
 ]
 location-groups: [
  {
   name: Japan
   locations: [
    Yokota AB
    Misawa AB
    Futenma MCAS
   ]
  }
  {
   name: Guam
   locations: [
    name: Andersen AFB
   ]
  }
 ]
 multipliers: [
  {
   asset-group: shelters
   location-group: Japan
   multiplier: 1.5
  }
  {
   asset-group: shelters
   location-group: Guam
   multiplier: 0.8
  }
 ]
}
```

Defaults Table

The Defaults table used by TAB-ROM is typically named *tblDefaults.csv* and describes the mitigations that must be included in each mitigation portfolio. Table A.6 presents more detail on the Defaults table. As Table A.6 shows, some of the rows and columns of the Defaults table are user-defined and must match items defined in other TAB-VAM input tables.

Table A.6. Defaults Table Description

Table Name	Defaults	
Example Path	tblDefaults.csv	
Table Description	The Defaults table lists the mitigations assigned to each location. This table contains a row for each location specified in TAB-VAM's Locations table and a column for each mitigation action modeled in the simulation. The values in this matrix specified the number of mitigation assets that must be included at each location in every TAB-ROM portfolio.	
Column Name[a]	**Data Type**	**Column Description**
Location	String	Must match the exact format of a location defined in TAB-VAM's Locations table.
A string name of a user-defined group of assets, such as the following: • Aeromedical evacuation • Aircraft decoys • Escort ships • Fuel storage defaults • Medical facilities • Missile defenses • Pipeline repair • Runway repair • Shelter defaults • Supply ships	Integer	The minimum number of the user-defined mitigation that must be assigned to this location.

[a] Most of the columns in the Defaults table are user-defined. Users can select assets and capabilities defined in TAB-VAM input tables as column names.

UpperBounds Table

The UpperBounds table used by TAB-ROM is typically named *tblUpperBounds.csv* and describes the maximum number of each mitigation that may be included in each mitigation portfolio at a given location. It is formatted identically to the Defaults table. Note that a user can decline to include locations or assets in the UpperBounds table that were in the Defaults table, but the user cannot include any locations or assets in the UpperBounds table that is not already included in the Defaults table. Table A.7 presents more detail on the UpperBounds table.

Table A.7. UpperBounds Table Description

Table Name	UpperBounds	
Example Path	tblUpperBounds.csv	
Table Description	The UpperBounds table lists the mitigations assigned to each location. This table contains a row for each location specified in TAB-VAM's Locations table and a column for each mitigation action modeled in the simulation. The values in this matrix specify the number of mitigation assets that must be included at each location in every TAB-ROM portfolio.	
Column Name[a]	**Data Type**	**Column Description**
Location	String	Must match the exact format of a location defined in TAB-VAM's Locations table.
A string name of a user-defined group of assets, such as the following: • Aeromedical evacuation • Aircraft decoys • Escort ships • Fuel storage defaults • Medical facilities • Missile defenses • Pipeline repair • Runway repair • Shelter defaults • Supply ships	Integer	The maximum number of the user defined mitigation that may be assigned to this location.

[a] Most of the columns in the UpperBounds table are user-defined. Users can select assets and capabilities defined in TAB-VAM input tables as column names.

Appendix B

Output Details

In this appendix, we describe the structures of the output tables that we described in Chapter 4.

Mitigation Portfolio Contents

Table B.1 provides more detail on the Chromosome-Gene table. Table B.2 provides more detail on the Gene table.

Table B.1. Chromosome-Gene Output Table Description

Table Name	Chromosome-Gene	
Example Path	Chromosome-Gene.csv	
Table Description	The Chromosome-Gene output table contains the list of genes in each chromosome	
Column Name	**Data Type**	**Column Description**
ChromosomeID	Integer	A unique identifier for each chromosome
GeneID	Integer	A unique identifier for each gene

Table B.2. Gene Output Table Description

Table Name	Gene	
Example Path	gene.csv	
Table Description	The Gene output table contains the list of all location-mitigation-quantity combinations used by TAB-ROM in the run	
Column Name	**Data Type**	**Column Description**
ChromosomeID	Integer	A unique identifier for each chromosome
GeneID	Integer	A unique identifier for each gene
Location	String	The location
There is one column for each mitigation in the Cost table provided as an input to TAB-ROM	Integer	The number of the mitigation specified by the column header at the location specified by the location in the Location column

Mitigation Portfolio Performance

Table B.3 provides more detail on the Chromosome table. Table B.4 provides more detail on the Chromosome ESRS table.

Table B.3. Chromosome Output Table Description

Table Name	Chromosome	
Example Path	Chromosome.csv	
Table Description	The Chromosome output table contains the cost and performance of each mitigation portfolio	
Column Name	**Data Type**	**Column Description**
ChromosomeID	Integer	A unique identifier for each chromosome
ElapsedTime	Double	The elapsed time, in hours, from when TAB-ROM started to when the chromosome was output. (This is normally the same for all chromosomes within a generation and allows users to monitor the time between generations.)
Cost	Double	The sum cost of all the mitigations added to the chromosome portfolio (i.e., what is added in addition to what was in the Defaults table input file)
EffectiveSortieRate	Double	The effective sorties rate is a metric calculated by TAB-VAM that includes all of the sorties that were able to leave the base, had sufficient aerial refueling (if applicable), and had a full munitions load (if applicable). This column will show the weighted average of the planned sortie generation achieved across the attack vectors, where any attack vector with a sortie generation more than 10 percentage points higher than the minimum performing vector must first have their value reset to be equal to 10 percentage points higher than the minimum performing vector. See the discussion surrounding Figure 2.1 in Chapter 2 or an illustration.
LocationSortieRate	Double	The location sorties rate is a metric calculated by TAB-VAM that includes all of the sorties that were able to leave the base, regardless of if they had sufficient aerial refueling or munitions to support their assigned mission. This column will show the weighted average percentage of the planned sortie generation achieved across the attack vectors, where any attack vector with a sortie generation more than 10 percentage points higher than the minimum performing vector first must have their value reset to be equal to 10 percentage points higher than the minimum performing vector. See the discussion surrounding Figure 2.1 in Chapter 2 for an illustration.

Table B.4. Chromosome ESRS Output Table Description

Table Name	Chromosome ESRS

Example Path	Chromosome-esrs.csv
Table Description	The Chromosome ESRS output table contains the effective sortie rate achieved for each attack vector for each chromosome.

Column Name	Data Type	Column Description
ChromosomeID	Integer	A unique identifier for each chromosome
Each of the user-defined attack vector names, as provided to TAB-ROM in the Allocation table input file, appears as a column header	Double	For the attack vector specified by the column header, the percentage of planned sorties that were able to leave the base, had sufficient aerial refueling (if applicable), and had a full munitions load (if applicable). The value in this table is the effective sortie rate for the allocation in the same row.

GA Data

Table B.5 provides more detail on the Chromosome Generation table.

Table B.5. Chromosome Generation Output Table Description

Table Name	Chromosome Generation
Example Path	Chromosome-Generation.csv
Table Description	The Chromosome-Gene output table contains the list of chromosomes in each generation along with the rank and crowding distance for each chromosome

Column Name	Data Type	Column Description
GenerationID	Integer	A unique identifier for each generation
ChromosomeID	Integer	A unique identifier for each chromosome
Rank	Double	A measure of the chromosome's performance relative to other chromosomes in the same generation. Nondominated chromosomes (i.e., those on the Pareto frontier) have a rank of 1. Chromosomes dominated only by rank-1 chromosomes have a rank of 2, chromosomes dominated only by rank-1 and rank-2 chromosomes have a rank of 3, and so on.
Crowding distance	Double	A measure of how "crowded" this chromosome's section of the Pareto frontier is. Chromosomes are considered more crowded if there are many other chromosomes that have similar cost-effectiveness values. Crowding distance applies only to the chromosomes on the current generation's Pareto frontier. For each objective (e.g., cost or effective sortie rate), a crowding distance is calculated for each chromosome. The

crowding distance is the ratio of (1) the difference in that objective's measure between its two immediate neighbors (one "better" and one "worse") to (2) the difference in that objective's measure between the extreme chromosomes on the frontier. This ratio is always between 0 and 1. These ratios are then summed for each chromosome to get its total crowding distance, which is therefore between 0 and *n*, where *n* is the number of active objectives.

Appendix C

Running TAB-ROM

In this appendix, we provide instructions for installing, setting up, and running TAB-ROM. We assume that TAB-VAM is already installed. Before attempting to run TAB-ROM, the user will need to understand how to and be able to run TAB-VAM.

Installation and Setup

Required Software

TAB-VAM is required to run TAB-ROM. Before running TAB-ROM, the user should ensure that their existing version of TAB-VAM runs with their current input setup.

TAB-VAM requires version 1.8.X or later version of the Java Runtime Environment, and the Parameter Sweep requires Python version 3.X. Both are available online at no cost. The most recent version of each should be installed prior to running TAB-ROM.

In addition, building TAB-ROM and/or TAB-VAM from the source code can be done using Apache Ant. We recommended using IntelliJ IDEA as the integrated development environment (IDE) while making code changes or building, but any Java IDE can be used.

System Requirements

TAB-VAM and TAB-ROM will run on any operating system. The computer's CPU must have at least one 2.0 GHz core with 4 GBs of RAM per thread, and 1 GB of free storage (10 GBs is preferred). Any number of cores can be used, but the amount of processing and RAM used throughout a TAB-ROM run is not consistent, so it is recommended to use fewer resources than what the system has in total.

Recommended Software

Many input tables are formatted as CSV files. Spreadsheet software, such as Microsoft Excel, will be useful for editing these tables. Inputs not contained in tables are stored in XML or HJSON. A text editor is needed to edit XML and HJSON documents. Notepad++ on Windows, BBEdit (formerly TextWrangler) on Mac, and Vim on Unix are excellent open-source text editors readily available online.

Outputs are saved in CSV tables. A good spreadsheet program (e.g., Microsoft Excel) or statistical analysis software (e.g., R) can be used to perform post-model analysis on the output files.

Installing TAB-VAM and TAB-ROM

The TAB-ROM package will include a compatible version of TAB-VAM and the Python scripts used to run each of them. Both models and scripts are unclassified. Users will be provided with inputs for both models. The inputs will come in the form of a directory that includes TAB-VAM and TAB-ROM input data files. They should include example input files for testing.

Users should copy the directory to the location on their computer where they wish to run TAB-VAM and TAB-ROM. Next, they should copy the TAB-VAM and TAB-ROM executable files and scripts into the top level of the provided directory. **By default, the inputs and outputs will be stored in the same directory where the model runs, and each new run will overwrite the outputs directory. Also, the active directories of TAB-VAM and TAB-ROM are different.** Generally, this only affects the files found in the resources folder; either these files will need to be moved one level up in the file system or the TAB-VAM default inputs file will need to be modified to reflect the relative location of the files from the active directory.

Running TAB-ROM

TAB-ROM is run from the command line. For a list of useful commands, refer to Table C.1.

How to Run TAB-ROM

TAB-ROM runs as a Java model contained in the ROM.jar file. This is run from the command line using a Python script or via a direct call to the Java virtual machine. TAB-ROM first checks to see if there is an existing set of seed files in the seed folder in the TAB-ROM active directory. If there is, TAB-ROM will resume the model run from the most recent completed generation in the seed file. Note that incomplete generations are not recorded in the seed files, and generations will not be saved in any way if the model is stopped prematurely. The seed files are optional, and if there is no existing set of seed files, TAB-ROM will generate a list of genes and gene instances based on the initial mitigations and upper bounds files. To directly call TAB-ROM from the command line, users can use the Java command that lists the ROM.jar as the jar file and use the option inputsDir to point to the preferred TAB-ROM inputs directory. An example of the command is as follows:

```
java -jar ROM.jar -inputsDir ../rom
```

Seeding TAB-ROM

As described above, users can resume a previously started TAB-ROM run by moving the output files generated by TAB-ROM to the seed folder in the inputs directory (specified in the Java command that calls TAB-ROM) and then by starting TAB-ROM as normal.

Users can also modify these seed files to do several things:

- Users can "pre-seed" specific genes or chromosomes into the existing outputs so they are used in the creation of the next generation of chromosomes when TAB-ROM is resumed. This pre-seeding can be done by editing the gene values in the most recent generation of chromosomes in the chromosome-gene.csv file. Users will also need to check if a chromosome is from the most recent generation by cross-referencing the chromosome-generation.csv file. Users may wish to pre-seed if they have a strong intuition about what the most fit genes are, or if they believe that TAB-ROM has become stuck in a local maxima or utility plateau and it is taking too long to escape or cross it.

- Users can add new genes that they did not include in the original TAB-ROM run. To add new genes, users first need to add them to the genes.csv file and assign them a unique gene ID. Users will also have to add the gene to the existing chromosomes in the chromosome-gene.csv file. Users may wish to add genes if they want to include new mitigations that were not in the original run to preserve what TAB-ROM learned while optimizing the existing chromosomes. Users should not attempt to remove existing genes in this manner because it significantly complicates the matter both from the standpoint of modifying the files and from an optimization standpoint.

Useful DOS and UNIX Commands

If users are not familiar with DOS or UNIX, the basic commands shown in Table C.1 will be useful. For the most part, the user interacts with TAB-ROM through the input and output files, not the command line. The four commands below should be all that users need to know to use TAB-ROM.

Table C.1. Useful DOS and UNIX Commands

Command	Description	Example Usage	Example Description
cd	Changes directory	cd C:\files	Uses the absolute file path to change the directory to C:\files
		cd myfiles	Uses the relative file path to change the directory to a subdirectory of the current directory called myfiles
		cd ..	Changes the directory to the parent of the current directory
Tab key	Autocompletes the current command	After typing "cd," hit the tab key	Autocompletes with a directory inside the current directory. On DOS, the tab key will autocomplete to the first item alphabetically and cycle through the list on each subsequent press of the tab key. On UNIX, the tab key will autocomplete to the point it needs to deviate. For example, if there are directories named "Angel" and "Andrew," DOS will autocomplete "A" to "Andrew" because it comes first alphabetically. UNIX will autocomplete it to "An" because it doesn't know if there should be a "d" or "g" next.

Command	Description	Example Usage	Example Description
		After typing "java -jar," hit the tab key	Autocompletes with a .jar file in the current directory. Continue hitting tab to cycle through all .jar files.
Up arrow	Repeats the previous command	Hit the up arrow	Reproduces the previous command, which can then be edited if needed. Continue hitting the up arrow to cycle through previous commands.
CTRL + C	Cancels the current process	While TAB-ROM is running, click anywhere in the command window and hold down the control and C keys simultaneously	The model will pause and ask the user if they want to cancel the process. Enter "y" to stop the run or enter "n" to continue.

Abbreviations

APO	Aimpoint Optimization
APT	Asset Placement Tool
CODE	Combat Operations in Denied Environments
CONOPS	concepts of operation
CSV	comma-separated values
ESRS	effective sortie rate
GA	genetic algorithm
GAD	Global Airfield Database
GF-VAM	Ground Forces Vulnerability Assessment Model
HJSON	Human JavaScript Object Notation
JSON	JavaScript Object Notation
Lean-START	Lean Strategic Tool for the Analysis of Required Transportation
NSGA	Nondominated Sorting Genetic Algorithm
OSD	Office of the Secretary of Defense
PRePO	Prepositioning Requirements Optimization
TAB-ROM	Theater Air Base Resiliency Optimization Model
TAB-VAM	Theater Air Base Vulnerability Assessment Model
USAF	U.S. Air Force
XML	extensible markup language

References

Boyd, Stephen, and Lieven Vandenberghe, *Convex Optimization*, Cambridge University Press, 2004.

Deb, Kalyanmoy, Amrit Pratap, Sameer Agarwal, and T. Meyarivan, "A Fast and Elitist Multiobjective Genetic Algorithm: NSGA-II," *IEEE Transactions on Evolutionary Computation*, Vol. 6, No. 2, April 2002.

Goldberg, David E., *Genetic Algorithms in Search, Optimization, and Machine Learning*, Addison-Wesley Pub. Co., 1989.

Gong, Wei, Qingyun Duan, Jianduo Li, Chen Wang, Zhenhua Di, Aizhong Ye, Chiyuan Miao, and Yongjiu Dai, "Multiobjective Adaptive Surrogate Modeling-Based Optimization for Parameter Estimation of Large, Complex Geophysical Models," *Water Resources Research*, Vol. 52, No. 3, March 2016.

Ishizaka, Alessio, and Philippe Nemery, *Multi-Criteria Decision Analysis: Methods and Software*, Wiley, 2013.

Lynch, Christopher, Rachel Costello, Jacob L. Heim, Andrew Karode, Patrick Mills, Robert S. Tripp, and Alan J. Vick, *Operational Imperative: Investing Wisely to Bolster U.S. Air Bases Against Chinese and Russian Attacks*, RAND Corporation, PE-A1996-1, January 2023. As of March 12, 2025: https://www.rand.org/pubs/perspectives/PEA1996-1.html

Markowitz, Harry M., *Portfolio Selection: Efficient Diversification of Investments*, Yale University Press, 1971.

Thomas, Brent, Mahyar A. Amouzegar, Rachel Costello, Robert A. Guffey, Andrew Karode, Christopher Lynch, Kristin F. Lynch, Ken Munson, Chad J. R. Ohlandt, Daniel M. Romano, Ricardo Sanchez, Robert S. Tripp, and Joseph Vesely, *Project AIR FORCE Modeling Capabilities for Support of Combat Operations in Denied Environments*, RAND Corporation, RR-427-AF, 2015. As of September 25, 2023: https://www.rand.org/pubs/research_reports/RR427.html

www.ingramcontent.com/pod-product-compliance
Ingram Content Group UK Ltd.
Pitfield, Milton Keynes, MK11 3LW, UK
UKHW051831310725
7180UKWH00014B/659